Searching for Human Origins

Story by **Jean-Loup Craipeau**
Factual accounts by **Pascal Picq**
Activities by **Béatrice Garel**
Game by **Catherine Pauwels**

BARRON'S

Contents

Story

Mega-infos

Game

Mega-infos

Anecdotes

Mega-infos

Activity

Mega-infos

Activity

Quiz

Stickers

Picture-Cards

Story

Yayna and Nom

Jean-Lóup Craipeau

The Cave of the Bear Clan

Yayna came instantly awake when she heard the sound. What was it? And was it inside or outside the cave? The night was unusually cold. Yana didn't want to push aside the furs that covered her sleeping place and go to investigate. But then the orphaned bear cub made a whimpering sound, she got up carefully so she wouldn't wake her

mother. Quietly, Yayna made her way down the rocky, dawn-lit path to where the cub was tethered.

"What is it, Urs?" she asked, brushing the fresh spring snow from his head and back.

Then she heard the sound again. First a sharp cracking, and, in the next instant, a roar like nothing she had ever heard before in all her eleven summers. She clung to the small bear as rocks and soil rained down on them.

When she finally dared look up, she saw that the entrance to the cave had disappeared. Forgetting her bruises, she ran back up the slope, calling to her mother. But even as she did she realized that she would never hear her mother's voice again . . . Or the voices of anyone in her clan. They were all gone, buried under the fallen roof of the cave. She tugged at the rocks until her fingers were bleeding. Then, cuddling the cub close to her, she let the tears come.

"We're both orphans now," she sobbed.

Yayna and the cub wandered for several moons, sharing fear, hunger, roots, and berries. As the sadness became less, sometimes they even took a few moments for a playful hiding game. Yayna kept hoping that they would see other people, but on the summery day when they did she almost wished they hadn't.

The two men who stood before her were gesturing angrily. One of them pushed her away, pointing back in the direction

from which she had come. But the other grabbed her arm and pulled her after him.

They came to a cave and Yayna and Urs were pushed inside. The fierce discussion between the two men continued. Finally, the one who wanted her to stay seemed to win the argument. An old woman showed her where she should sleep. Urs snuggled down beside her. Before her eyes closed, Yayna glanced toward the opposite wall of the cave. There she saw a boy about her own number of years. He looked back at her for a long time and Yayna knew they would be friends.

The boy's name was Nom. As the weeks passed, he taught Yayna how to chip and shape stones and how to skin a deer. Best of all, he taught her how to make beautiful sounds by blowing into a reed pipe with holes cut in its side.

It was Nom's father who had argued that she be allowed to stay with them.

"You see," he often said, "there is no need to fear others just because they are different. Besides, we are the Bear Clan and she has a strong bear spirit. Look how he follows and protects her."

Yayna didn't know it, but she was an early human of a kind we now call Cro-Magnon. Nom and the people who gave her a new home were the kind of early humans we call Neanderthal.

Story

The Great Woolly Rhino Hunt

One day Nom asked Yayna to go hunting with him. He had a plan for trapping a woolly rhino. Behind a hill he would start a fire. The burning grass would drive the rhino around the hill and down a slope to where Yayna would be waiting in front of a large ditch. Yayna was so quick and agile that Nom knew she would be able to jump out of the rhino's way. But the huge beast would not be able to change direction in time and would plunge into the ditch.

Yayna was there, listening for the sound of the rhino's hooves, when she was startled by a very different noise. Quite close behind her she heard an angry grunt. She turned to see three men coming toward her. They looked like the people in Nom's family, but she knew they must be from a different clan.

"The beast is ours!" one of them said. "We have been trailing it for three suns!"

Yayna put the music pipe to her lips and blew the sound she and Now used to call each other. Maybe, she hoped, the notes would even bring Urs.

The second of the men called her by the name she had heard some of the people of the Bear Clan say behind her back.

"No," she said, "I am Yayna, of the Bear Clan."

"And I am Krok, of the Deer Clan," said the biggest of the three. "I spill the blood of the likes of you."

"You have to catch me first," Yayna said, leaping quickly up onto a nearby boulder.

Yayna sensed that Urs was close. When he came running, growling fiercely, the three men turned and fled.

"Good Urs," Yayna said, jumping down from the rock and ruffling the bear's furry back. He rubbed his huge head against her.

"Ai," she said, "Careful. You're no longer a little cub."

Then there it was, the thundering hooves of the running rhino! Yayna stood in front of the ditch. She knew that Urs would have sense enough to get out of the big beast's way.

The rhino appeared suddenly, charging straight toward them. Urs quickly moved away. At the last second, as Yayna felt the beast's hot breath upon her, she leaped back up onto the boulder. The earth shook as the rhino fell into the deep ditch.

"Ai—eeee !"

Yayna heard Nom's cry and soon he was beside her, his dark eyes bright with excitement.

"The clan will have plenty of meat to keep in the snow," he said.

"And a beautiful horn too," Yayna said. "Unless the men of the Deer Clan come back and take the beast from us."

When Yayna mentioned the Deer Clan men, she saw that Nom was troubled by more than their threats to take the rhino.

"They should not be here," he said. "It means their own hunting grounds have shrunk because . . . "

"Because of my people," Yayna said, finishing his sentence. "They are coming into your lands. I understand now why some of your people dislike me."

"It is not your fault, Yayna. Anyway, you are one of us now. And I . . . I do not dislike you."

He held Yayna close to him, causing her heart to pound harder than it had when she jumped away from the charging rhinoceros.

"But now you must run, Yayna," Nom said. "Take Urs with you so you will be safe. Tell the men that the rhino is here for them to kill. I will stay and watch."

"The Spirit of the Bear Has Left My Son"

When Yayna, running ahead of the others, returned to the ditch, she found Nom lying at its edge, bloody and unconscious. She knelt beside him and looked around wildly. She saw that one of the three Deer Clan men was lying nearby and that another had fallen into the ditch, where he had been trampled by the raging rhinoceros. The third one was gone.

Yayna stroked Nom's hair. Urs licked his face. The Bear Clan medicine man arrived and tended to his wounds. Still, Nom did not move. His father stood nearby, his head bowed in grief.

Finally he said, "The spirit of the bear has left my son. I will carry his body home."

Yayna looked up to see the third Deer Clan man returning. Many of his people were with him, women and children among them, gaunt with hunger.

The father of Nom spoke again. "We will leave the rhino," he said.

Yayna understood then why he was called the Wise One.

They began the walk back to the cave, Nom in his father's arms. Yayna, walking beside them, held Nom's hand.

"Oh, Nom," she murmured, "without you nothing will be the same. I know you died bravely and that you will travel in the big sleep with the skull of a bear and ten ibex horns. But Urs and I—we will never forget you. I love you, Nom."

The hand she was holding squeezed back. And then Nom opened his eyes and smiled weakly.

"Would you say that again?" he asked.

"Oh, Nom, you're alive!" Yayna cried. "Yes, yes . . . I said . . . I love you !"

It took many suns for Nom's wounds to heal. When his son was well again, Nom's father asked everyone in the clan to listen to what he had to say.

"Yayna and Nom have taught us much," he said. "It is time for them to have their own hearth, their own space in the cave."

We can only hope that Yayna and Nom lived happily ever after.

The Family of Man

We don't know all the answers, but we are learning more about human origins all the time. The story of Yayna and Nom is fiction—but it does offer an idea of what the lives of early people may have been like.

In the Beginning . . .

Most scientists agree that, about 12 billion years ago, the universe came into being in a cosmic explosion we call the "Big Bang." Our solar system is about 4. 5 billion years old. The ancestors of modern man did not appear until much later, probably about 3 million years ago.

Man

Australopithecus robustus

Australopithecines

The Evolution of Human Beings

Some of the branches of our family tree died off, as in the case of *Australopithecus robustus.* But *Australopithecus africanus* led finally to *Homo sapiens,* ancestors to today's humans.

Mega-infos

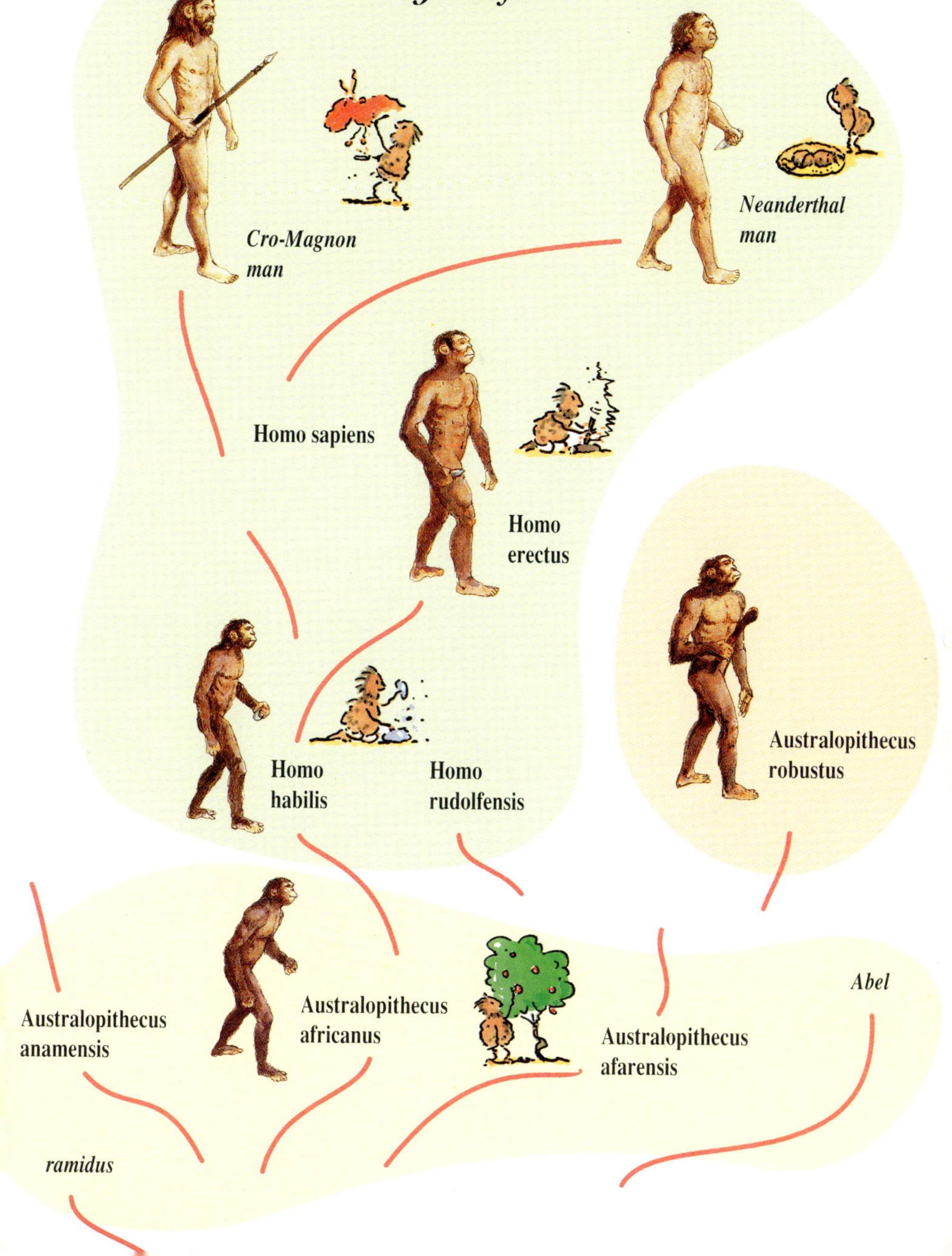

Cro-Magnon man
Neanderthal man
Homo sapiens
Homo erectus
Australopithecus robustus
Homo habilis
Homo rudolfensis
Australopithecus anamensis
Australopithecus africanus
Australopithecus afarensis
Abel
ramidus

The Amazing

The history of people began somewhere in east Africa, around 7 million years ago.

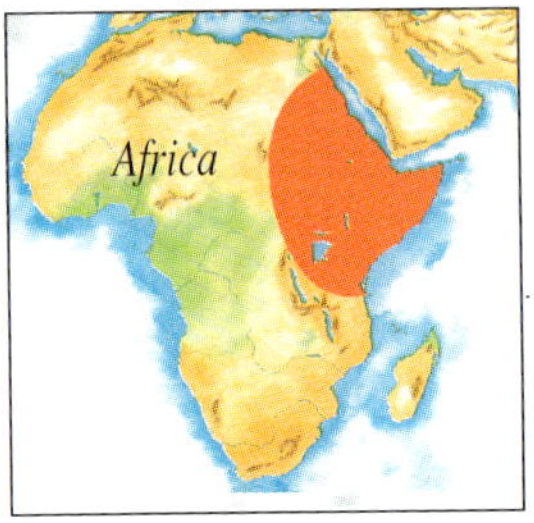

7 Million years ago . . .

. . . large monkeys developed. They lived in the forests and on the savannahs.

Between 4 and 3 million years ago . . .

. . . the ancestors of modern man, called *hominids*, spread all across Africa. They inhabited most areas of the continent—with the exception of the tropical forests.

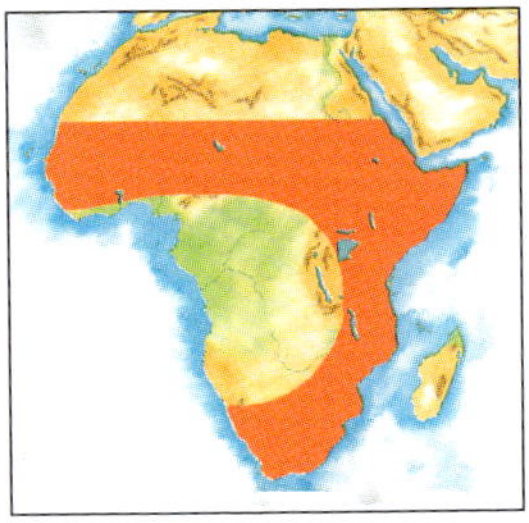

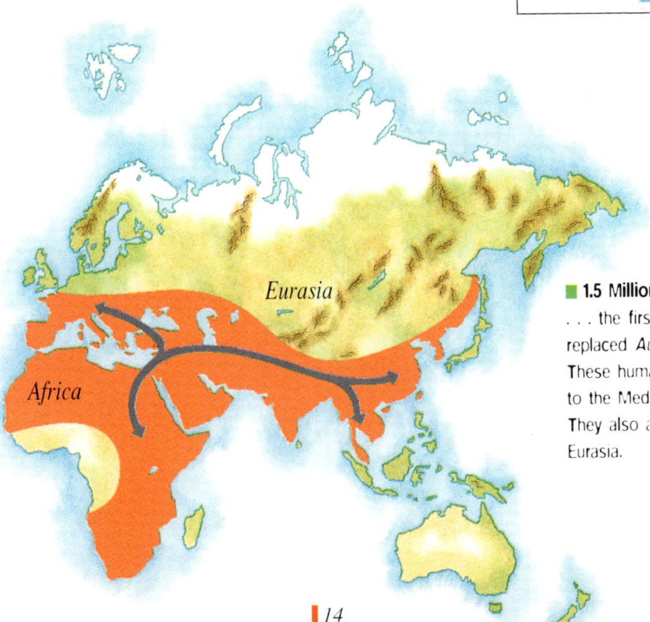

1.5 Million years ago . . .

. . . the first men had replaced *Australopithecus*. These humans spread north to the Mediterranean Sea. They also appeared in Eurasia.

Story of People

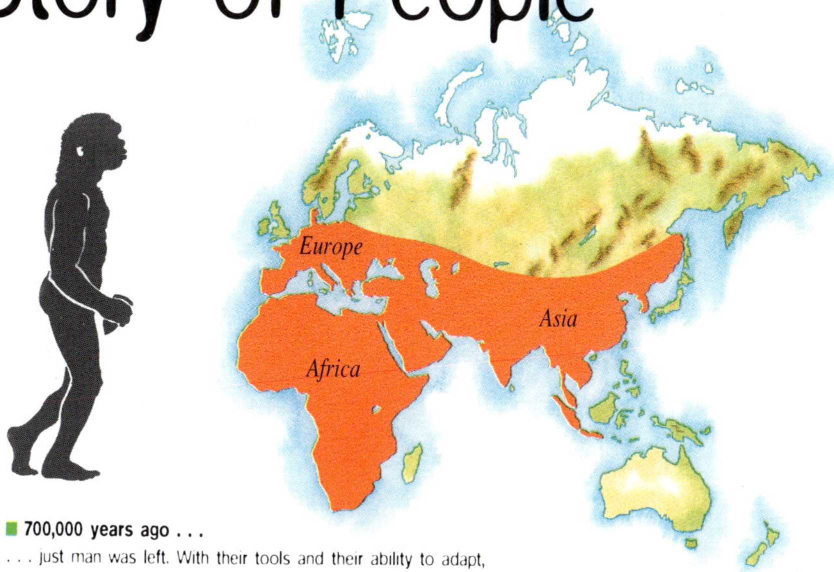

700,000 years ago . . .

. . . just man was left. With their tools and their ability to adapt, humans spread to all of Africa, and to large parts of Europe and Asia.

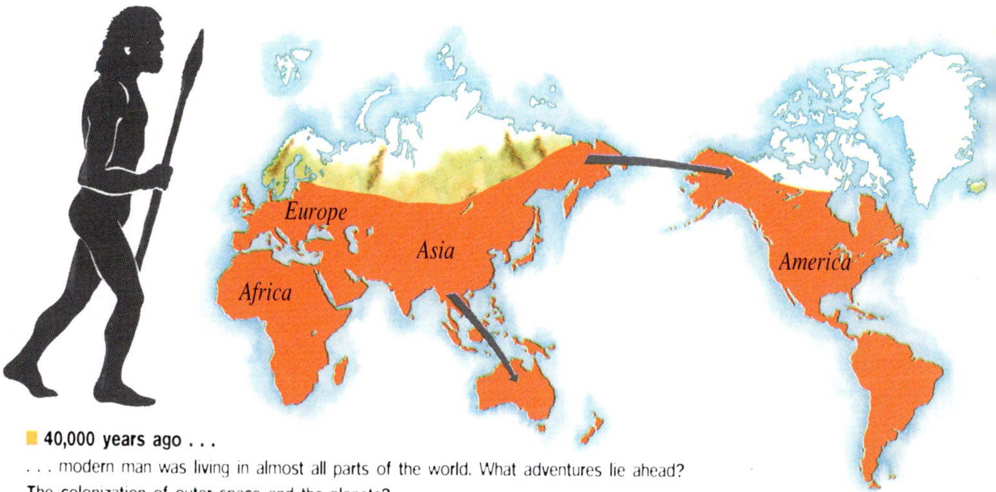

40,000 years ago . . .

. . . modern man was living in almost all parts of the world. What adventures lie ahead? The colonization of outer space and the planets?

Man and

Man is not a direct descendant of the great apes. The two species separated 6 to 8 million years ago.

Genes

All the cells in our bodies contain special, microscopic particles. Called genes, they are responsible for the similarities we have to our parents and to our relatives. Each gene controls the development of some characteristic of the offspring.

Genes and Generations

Genes make us what we are. Yours were passed on to you by your parents. You, in turn, will pass them on to your children.

The Story Told by Our Genes

Our genes contain a substance called *DNA*. Our DNA reveals that humans and today's apes have the same ancestors.

the Apes

Distant Cousins

An astonishing 99 percent of human genetic structure is identical to that of gorillas and chimpanzees.

Monkey Business

Our behavior is strikingly similar to that of chimpanzees, the smartest of the apes. We both use tools, are excellent hunters, eat some of the same foods, live in organized groups, and sometimes go to war.

chimpanzee

gorilla

What Is Evolution?

Evolution is the process by which nature gradually changes animals, plants, insects, fish, viruses—all living things.

Survival of the Fittest

When the environment changes, some species die out. Those that remain do so because they were born with characteristics that better fit them to the new situation. It might be color, size, or shape of a body part. Through their genes, the survivors pass these advantages on to their offspring. This process is called natural selection.

Major Changes

Most changes are slight and occur over the course of several generations. Sometimes, however, animals or plants change so much that they become a new species. This is what happened in the evolution of humans.

Natural Disasters and Evolution

70 million years ago, dinosaurs ruled the Earth. Then a natural disaster—perhaps a huge comet colliding with Earth—wiped them out. The small animals that had lived alongside the giant beasts were able to develop without danger of being eaten or trampled to death. These were the first mammals, the creatures that eventually evolved into apes and humans.

This valley contains a small lake and is dominated by a volcano.

The volcano erupts and the lava flow changes the course of streams. The lake grows larger.

The climate becomes more humid. Forests cover the mountain slopes and provide homes for woodland creatures.

The Development of the Hominids

The earliest hominid fossils we have found are between 5 and 8 million years old. They were discovered in east Africa.

The Great Rift Valley

Between 5 million and 7 million years ago, the continent of Africa changed dramatically. The Rift Valley was being formed—more than 3,000 miles (4,800 km) long, 50 miles (80 km) wide, and in places more than a mile (1. 5 km) deep. The valley brought about changes in the climate. On its west side there were tropical forests, while to the east the forests were largely replaced by dry, grassy savannahs.

Easy Living for the Apes

The great apes continued to live comfortably in the forests west of the Rift Valley. There they found plenty of food and water. Because conditions were so favorable, they were not forced to change.

The Struggle of the Hominids

The hominids east of the Rift Valley had to adapt to a more difficult way of life. Because the woodlands were few and far between, they traveled long distances. Because they had to hunt for food on the savannahs, hominids that could walk and run upright had an advantage. They survived and passed on this adaptation to their offspring.

Africa, the Cradle of Man

The oldest fossils yet discovered have been found east of the Rift Valley. Some of these fossils are very puzzling. Are they the remains of apes or early hominids?

Lucy—the

Lucy, the oldest known hominid ever found, lived about 3,000,000 years ago.

Who Was Lucy?

Lucy was a young female *Australopithecus afarensis* whose bones were found in the *Afar triangle*❦. She was about 18 years old at the time of her death. The scientists who discovered her named her Lucy because, on the day she was found, they were playing a tape of the song "Lucy in the Sky with Diamonds."

❦***Afar triangle***
A region in Ethiopia.

What Was Lucy?

Lucy was neither monkey nor human, but something in between. Her ribcage and arms were more like those of a chimpanzee, but her lower body was similar to ours.

Most Famous Fossil

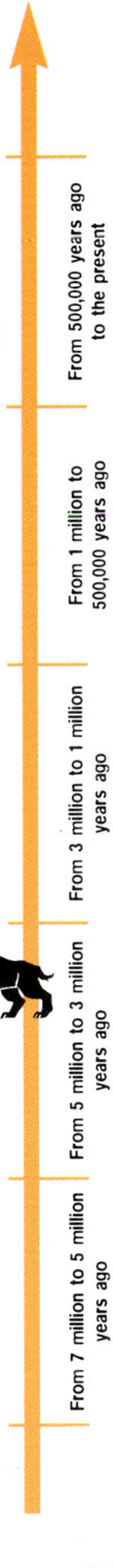

Lucy's Family

When Lucy was discovered, in 1974, she was the only one of her kind ever found. Within a year, remains of thirteen other individuals—men, women, and children—had been located. They became known as the "first family."

Lucy's Life

Like others of her kind, Lucy lived in a group. Her curved toes and finger bones told scientists that she spent much of her time in the trees. However, other bodily characteristics indicated that she also walked upright. There is evidence that Lucy and her family ate meat and *tubers*❦.

❦ ***Tuber***
The fleshy underground part of certain plants.

Lucy's World

Lucy and other *Australopithecus afarensis* lived in dense forests edged by grassland. They shared the trees with many kinds of monkeys. Living on the grassland were antelope, buffalo, horses, and elephants. Preying on these were saber-toothed tigers, lions, and hyenas. Probably the meat Lucy's family ate was killed by these predators.

The Changing Hominids

By about 2.5 million years ago, big changes had come to east Africa. As the savannahs replaced much of the forest, hominids and the first creatures we call human replaced the australopithecines.

Living Side by Side

For a long time the hominids we call "near men" lived in the forests and savannahs of east Africa. At the same time and place were new hominids developing into the family of man.

A New Lifestyle

The hominids we call *Homo habilis* and *Homo rudolfensis* still lived close to the trees in the forest that remained. As time went by they started to move out onto the grassy savannahs. There they found a variety of edible plants. They had no weapons for hunting, so they became scavengers, eating the meat left behind by predators.

Scavenger
An eater of dead plant or animal matter.

Handy Man

Early men, and the australopithecines who lived side by side with them, faced many challenges on the savannahs. Digging in the ground with bare hands was hard, so they began to use sticks and bones as tools in their search for edible roots. Getting meat off the carcasses they found was difficult until the men we call *Homo habilis* ("handy man") began using sharp stones to carve up the meat. As long ago as 1.9 million years, *homo habilis* were chipping away, fashioning basalt and jasper rocks into choppers, scrapers, and hammer stones.

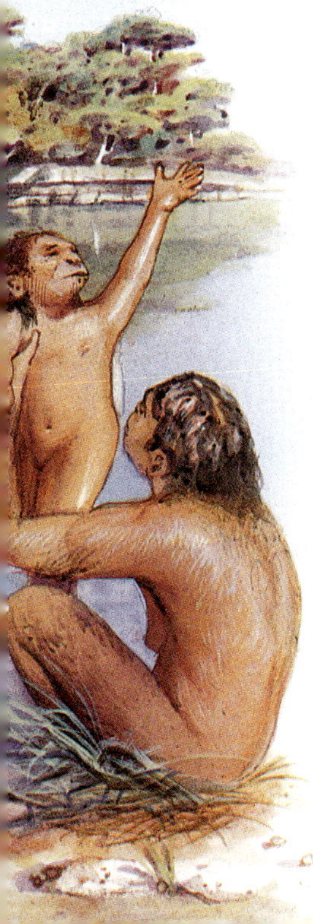

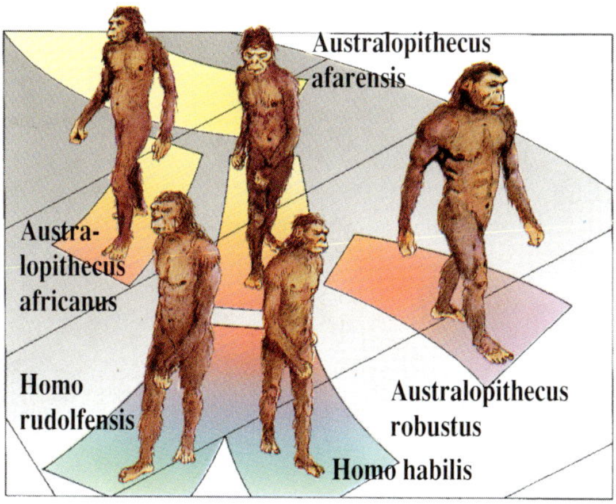

More Like Us

Different foods and ways of life brought changes in the hominids' bodies. Their brains got larger, their jaws and teeth smaller. They began to look more like humans of today.

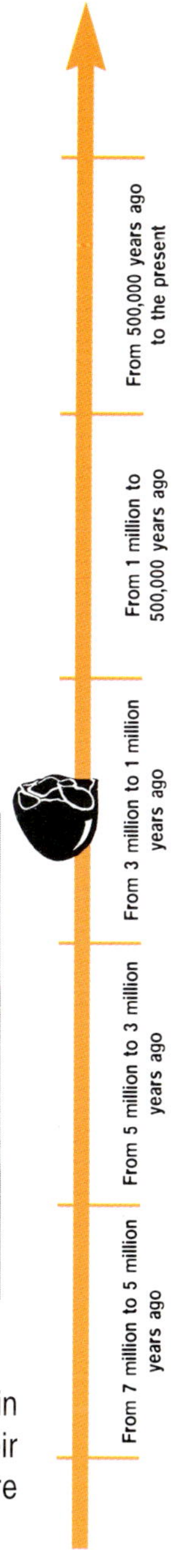

Man Takes Over

Hominids that could not adapt to the changing climate disappeared. Only man was left.

The First Tall Human Beings

Do you think of early humans as being short? Wrong. *Homo erectus,* who appeared about 1.6 million years ago, was tall—perhaps as much as 6 feet (almost 2 meters)—and weighed about 160 pounds (72.5 kilograms).

Man, the Runner

Running across the savannahs gradually led to changes in the bodies of *Homo erectus.* With long legs, barrel-shaped rib cages, and big shoulders, their fossilized skeletons resemble those of modern man.

Homo erectus, the Inventor

Homo erectus, with his bigger brain and skillful hands, became an inventor. He made sharp spearheads that helped him bring down big animals. He learned how to use fire. He built shelters with tree branches. He could go on longer hunts, get more food, and live more comfortably.

Homo erectus, the Survivor

Many kinds of early man died out, but *Homo erectus* was around for a long time, from about 1. 6 million to about 200,000 years ago. He was helped by his weapons, tools, more meat in his diet, and especially by his bigger brain.

Toward New Lands

With his ability to live off the land, adventurous *Homo erectus* could move to new locations. Some migrated to Europe and Asia. They were able to adapt to different climates and climatic changes that had wiped out other kinds of early man.

Where Do the

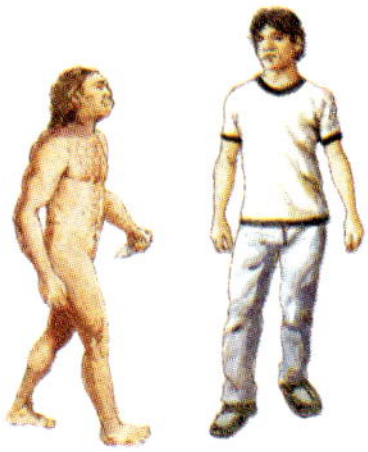

During Earth's last Ice Age❦, glaciers made life difficult for Homo erectus. The more robust Homo neanderthalensis succeeded them.

❦ *Last Ice Age*
The period between 75,000 and 10,000 years ago. Glaciers covered much of the modern temperate zone.

Who Were the Neanderthals?

Neanderthals first appeared about 300,000 years ago. By 30,000 years ago they had disappeared. They were powerfully built, probably two or three times stronger than people today. Their brains, on average, were larger than our own.

Neanderthals Fit In?

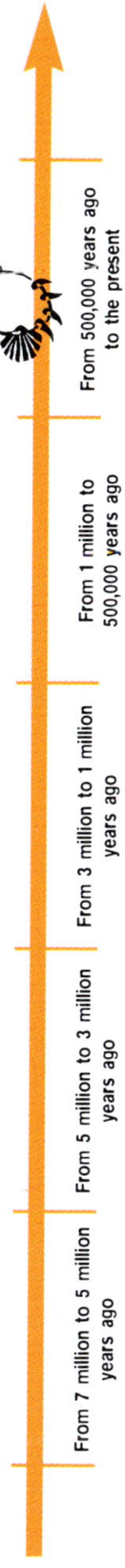

The Wandering Neanderthals

The Neanderthals were wanderers, traveling in and out of the regions where they first appeared. Their remains have been found in Europe, the Middle East, and central Asia. During periods of migration they encountered ancestors of Cro-Magnon man. Neanderthals and Cro-Magnons flourished together for more than 60,000 years. Did they compete for food? Did they mate with each other and have children? We may never know.

The Prehistoric Golden Age

The Neanderthals shaped stones into tools. Their skill as hunters and the abundance of game gave them time to care for the sick and the elderly and even to make ornaments and carvings, some of which have been found alongside Neanderthal fossils.

Neanderthal Beliefs

The Neanderthals may have been the first people to develop a religion. They buried their dead, laying out their bodies on ground carpeted with flowers. Did they believe that an individual had a spirit that survived death? It seems quite possible that they did.

The Origin of

The first hominids with completely modern characteristics are called Cro-Magnon. They spread all over the world, while other hominids became extinct.

Homo sapiens—*from the Latin word for* wise. *The only surviving hominid, modern man.*

The Final Survivors

From 35,000 years ago to the present there has been only one type of hominid: *Homo sapiens*. This adaptable human was able to cope with a range of changing environments. Eventually, they replaced all other human populations.

Where did People Come From?

Modern people appeared around the middle of the last ice age. Where did they come from? Did they come out of Africa or Asia and emigrate to Europe? Some fossil finds support this idea. The origins of modern people are not known with certainty.

Modern People

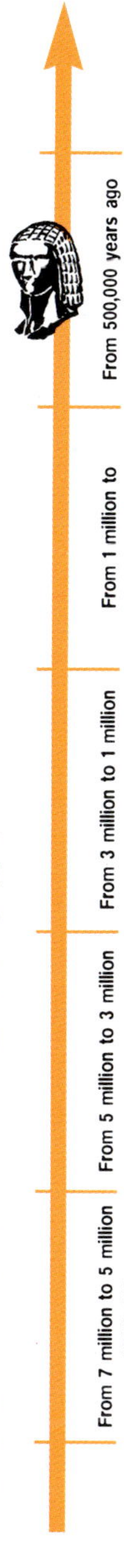

A leather thong helped a lance go faster and farther.

The Great Migrations

Homo sapiens migrated to all parts of the globe. They settled in the forests of Africa and Indonesia. They were able to live on the coldest steppes and in the hottest deserts. By 50,000 years ago they were in Australia, and around 30,000 years ago they began to move into the Americas. These capable humans flourished anywhere their spirit moved them.

The Advance of Modern Man

Modern humans developed new and better tools. With the start of farming, they began to live in settled communities. They fashioned jewelry, made sculptures, and engaged in *bartering*. Humans began to change culturally rather than biologically.

Bartering
Exchanging one item for another.

One Day Is Just Like Another

Is this a glimpse of life on an ordinary day 30,000 years ago? Not exactly. The artist has drawn 19 objects or activities that played no part in prehistoric lives. Can you find them?

Answers on page 63.

Game

Science Detectives

Archaeologists study tools and other objects made by people.

Paleoanthropologists *search for fossilized skeletons of hominids.*

Paleontologists *search for and study the fossils of plants and animals.*

Sedimentologists *analyze the sand and mud deposited by the wind and water.*

Geologists *study soil, rocks, and land formations.*

Geochemists *study the chemistry of soil, rocks, and fossils.*

Palynologists *analyze the pollen of ancient plants.*

It takes a lot of scientific teamwork and many different kinds of scientists to solve the mysteries of man's past.

Layers of Time

Wind and water deposit sediments on the Earth's surface. Layers of sediment become compressed into stone as they pile up, one on top of the other. Those that contain evidence of man's presence at the time of their formation are called archaeological layers.

at Work

Looking for Clues

Many clues about man's past are found in archaeological layers. The location of every little bone or piece of bone, of every tool or fragment of tool, is carefully recorded. Bits of pollen provide information about the plants of a given time period and reveal the time of year the site was occupied. Bones and remains of animals show what the people who lived at the site were hunting and eating.

Secrets in the Soil

Sometimes fossils are found lying on the ground. Usually, however, scientists have to dig to turn up traces of early humans. Since they know the age of the archaeological layer in which they are digging, they know the age of the fossils and objects found in it. By studying these remains they learn much about the lives and activities of species who were alive when that layer was forming.

An archaeological site is carefully measured and marked off with ropes to form a grid.

Amazing but True!

Footprints from the Past . . .

In 1976, in Laetoli, Tanzania, some playful paleontologists were throwing dried, pancake-shaped elephant droppings at each other. Ducking to avoid one of the huge, hard missiles, one of the scientists fell and found his nose in an ancient footprint! What he had discovered was a track left by an upright hominid who had walked there 3.5 million years before. The ground at that time had been covered by soft volcanic ash that hardened soon after the walker had passed.

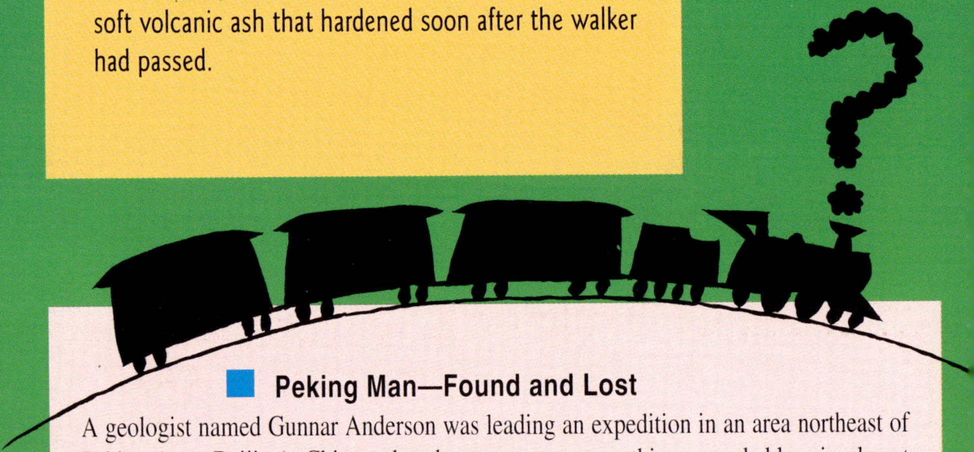

Peking Man—Found and Lost

A geologist named Gunnar Anderson was leading an expedition in an area northeast of Peking (now Beijing), China, when he came upon something remarkable: six almost complete skulls and the remains of forty individuals. Before the 360,000-year-old fossils could be shipped out of China, World War II had started and the Japanese had invaded China. The fossils were put on a train but it never reached its destination. Was it seized by the Japanese? Was the precious cargo lost or destroyed? Might it be found again someday? Fortunately, the expedition had made casts of the fossils so scientists have been able to take the measure of these ancient Asians we call Peking Man.

TAUNG'S CHILD

Raymond Dart, a professor of anatomy in South Africa, had always been fascinated by fossils. In a collection sent to him in the early 1920s, he made an exciting discovery. At first it looked like the skull of a young monkey, but Dart was able to determine that it had belonged to a being that walked upright. It was actually the skull of a young child with its "baby teeth" still intact. It had come from a quarry at Taung, in South Africa, so it has become known as "Taung's child." But Dart, certain that it represented a newly discovered species, gave it a more meaningful name: *Australopithecus africanus*. As the illustration on page 13 shows, *Australopithecus africanus* did not die off like *Australopithecus robustus*. Rather, it is believed to have started the branch of the human family tree that led to *Homo sapiens*—modern man.

HOW A DOG DISCOVERED A TREASURE . . .

One day in 1940 two teenagers and a dog were taking a walk in the countryside of southern France. The dog, running ahead, suddenly disappeared. The young people ran to see what had happened and ended up beside the dog in a hole that led to an underground cave. They looked around in amazement at scores of colorful action paintings of horses, bison, bulls, and other animals. Thus was discovered the famous Lascaux Cave. Its walls were decorated by primitive artists more than 15,000 years ago.

A Better Way

Early humans walked upright, on two legs. And that, as much as anything else, is why people made the advances they did.

The Origins of Bipeds

Seven million years ago, different kinds of hominids moved about in different ways. Some still lived in trees, so they were excellent climbers. Some walked about on all four feet. But some had become *bipeds*❦. They walked upright, with their hands freed for other activities. This brought about great changes in both physical and cultural development.

❦ ***Biped***
Any animal that walks on two feet.

to Walk

Feet for Walking, Hands for Holding

Sometime between 4 million and 7 million years ago, *Homo habilis* became a full-time biped. He found many advantages to being upright. He could use his hands to carry meat back from the kill. He could carry weapons and even use them on the run.

The Supreme Biped

People's body shape changed. They became leaner and taller, their legs longer. They could run fast, for long distances. The savannahs on which they hunted were hot and dry, so with each generation individuals had less body hair. Sweat glands developed to help their bodies stay cool. They were becoming more and more like people of today.

Homo erectus

Neanderthal man

Cro-Magnon man

Humans

Mastery of fire brought many changes to the lives of early humans. It gave them light, warmth, a means of cooking food, and protection from predators.

Starting a Fire

and Fire

People Find Fire

Early humans often saw flames leaping from a lightning-struck tree or in a grass fire racing across the savannah. At some time, perhaps a million years ago, a great idea dawned. Why not take some of the embers from the dying fire and use them to start a campfire?

Making Their Own Fire

Eventually people found ways to make fire for themselves. By rubbing two hard stones together they could produce sparks. By twirling a stick against a piece of wood they could make it hot enough to set fire to dry leaves or wood shavings. Now they could have fire whenever they wished.

The Tell-tale Hearths

Early man made fires more useful by building them in circles of stones that kept them from spreading. Stone hearths found in many parts of Europe and Asia tell anthropologists how *Homo sapiens* of long ago used fire.

A hearth

What Did They Eat?

Finding enough food was not always easy. Hominids learned to be omnivores❦, which helped them adapt to changing conditions and environments.

Food, Teeth, and Jaws

The jaws and teeth of australopithecus and "near men" developed the way they did because of the often-tough roots, vegetables, fruits, and nuts they ate. Strong teeth and jaws were needed to chew them. Anthropologists nicknamed some of these hominids "nutcrackers" because of their big jaws.

❦ ***Omnivore***
One who eats all kinds of food—meat, fish, fruit, tubers, etc.

Meat Made a Difference

About 2.4 million years ago, *Homo habilis,* who developed better ways of hunting, began to eat more meat. Meat was easier to chew than tough roots and some kinds of vegetation. The size of jaws and teeth, and therefore faces, gradually became smaller.

The First Backyard Barbecues

For a long time early humans ate all food raw. It was around 500,000 years ago that people started roasting vegetables in hot ashes and meat over open fires.

Cooking in Water

It took longer for our ancestors to catch on to the fact that cooking tough plant foods in hot water would soften them. By stitching hides together they made bags, filled them with water, and then dropped in stones that had been heated in fire. Do you suppose that at some time they added meat to the vegetables and made the first stew?

The Beginnings

People took a great leap forward when they learned to live in groups. A band of men working together had a much better chance of killing a large animal than one man hunting alone.

# of Society

A New Way of Life

People today enjoy gathering together around a campfire. So, probably, did the earliest humans. Group hunting and the mastery of fire changed the way they lived. Not only did fire give them protection from predators and a way of keeping warm, it gave them a feeling of community. One fire could warm many individuals. Perhaps they sat around in the evening and talked about the day's hunt.

The Growth of Language

The use of language increased because people in groups had more need to communicate with each other.

Women and Children

In group living, different individuals had different responsibilities. Women had more time to care for their children. Anthropologists have determined that our ancient grandmothers gave birth to a child every 3 to 5 years and that children stayed close to their mothers until the age of 10 or 11.

How Long Did Early Man Live?

Group living may have increased life spans somewhat, but primitive people did not live very long. Fossils tell us that both men and women were old by the age of 30 and didn't live much beyond that.

The

The early hominids and the people that followed them lived in many kinds of shelters.

Nests in Trees

Australopithecus often slept in the treetops, building nests of leaves and moss. There they were comfortable, warm, and safe from the predators that roamed the night forest. Later species used tree branches in a different way, collecting them to build huts.

Archaeologists have found remains of 2-million-year-old huts made from branches.

A Ready-made House: A Cave

A cave with a southern exposure can be a dry, warm home. Early man often lived in such natural shelters. Hides hung at the entrance kept out the cold and wind. Sometimes many family groups shared a large cave.

First Shelters

Shelters to Go

When they were near a forest, *Homo erectus* built huts of branches, covering them with leaves. When they traveled across the savannahs, where branches were scarce, they used huge mammoth bones for a framework and covered it with animal hides. The hides were all they had to carry with them when they moved on. When they stopped and camped they could always find more bones and quickly put up another shelter.

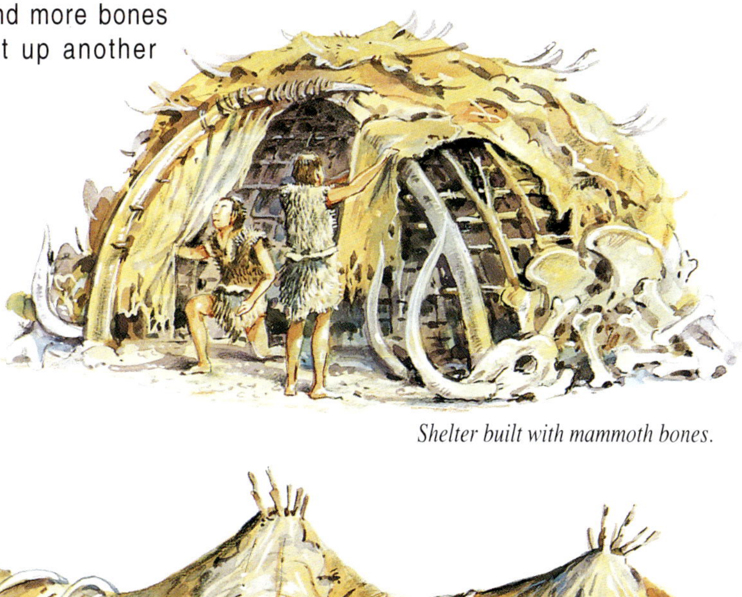

Shelter built with mammoth bones.

Tent made of animal hides.

Activity

You Can Build

At an archaeological site in the city of Nice, in France, archaeologists found the remains of shelters built by *Homo erectus* almost 400,000 years ago. They were made of branches braced by stones. Some were large enough to hold as many as 20 people.

Make a model of the *Homo erectus* hut

You will need:

- several small twigs, forked at one end
- modeling clay
- a piece of cardboard, about 8 × 12 inches (20 × 30 cm)
- some small pebbles
- some sand

1. Flatten the modeling clay so that it covers the piece of cardboard.

2. Near one of the 8-inch (20-cm) sides, stick two twigs into the clay to mark the entrance. Lean them slightly inward, forked ends at the top.

3. Continue placing twigs behind the first two. Each one should be a little shorter than the one in front of it. Link the forked ends together at the top by turning the twigs this way and that.

4. Curve the sides inward as you go and, when they meet at the back, place pebbles at the bottoms of the twigs to hold them in place. Cover the clay base with sand.

Activity

a Primitive Hut

If you live where you can find branches and rocks, you can make a lifesize hut.

Make a lifesize hut

You will need:

- 30 or 40 forked branches about 4 to 6 feet (1. 5 to 2 m) long
- 50 or more rocks

1. Build your hut on ground that is soft enough so that you can push the branches into the earth.

2. Put two branches in place to make the sides of the entrance, forked ends up.

3. Bring the tops together, turning the branches to link them.

4. Continue forming the sides in the way described in the instructions for building a model.

5. When the curved sides meet at the back, place rocks all around the base to hold the branches in place.

Man, the Successful

All hominids ate meat, but only humans learned how to hunt.

Hunter

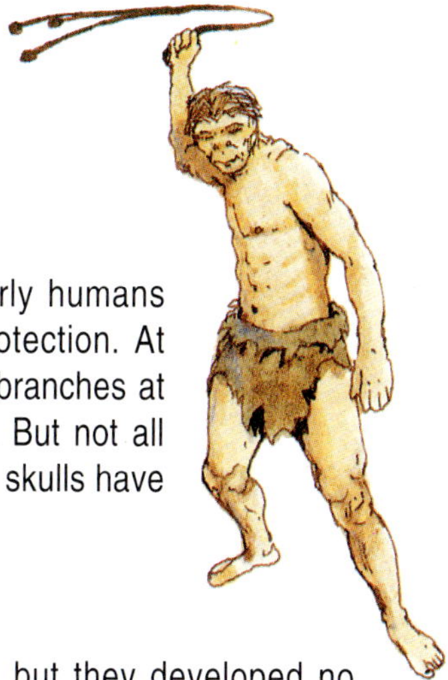

Before They Were Hunters

Long before they became hunters themselves, early humans were hunted. Living in groups gave them some protection. At night they built big fires and probably threw burning branches at the animals attracted by the smell of roasting meat. But not all people escaped the beasts that stalked them. Human skulls have been found that were scraped by teeth and claws.

The First Hunters

Australopithecus hunted, but they developed no weapons. They hurled sticks and stones to kill small animals. Sometimes they were able to bring down a small antelope. The only big animals they were able to kill were those that were injured or ill.

Man Becomes a Real Hunter

Homo ergastus and *Homo erectus* became the first real hunters. They were good runners, able to track prey over long distances. Hunting in groups, they drove rhinos, bison, horses, and deer over cliffs or into swamps. After killing an animal, they worked together to carry it back to camp. The abundance of meat in their diet helped make them stronger and smarter.

New and Better Weapons

Around 100,000 years ago, Cro-Magnon people invented throwing weapons. One of them, a device with stones fastened to leather thongs, could entangle the legs of prey such as antelope. Around 40,000 years ago, Cro-Magnon men made harpoons and used them to catch fish. Spears helped them bring down huge woolly mammoths. Bows and arrows came into use much later, only about 10,000 years ago.

Tools and More Tools

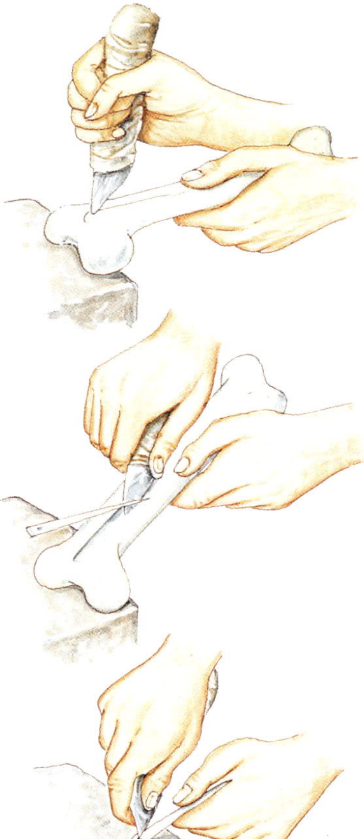

It took hominids 3 million years to progress from simply shaping stones to both shaping and polishing them.

The First Tools

Australopithecus used rocks as hammers for cracking nuts and breaking up hard roots. For digging up roots and tubers, which were abundant in the soil of the savannahs, they used large animal bones. It took early men a long time to think of using rocks and bones to make other tools.

Tools that Cut

Between 2 and 2.5 million years ago, an early man chipped away at a piece of hard rock, known to us as flint, and made it into a sharp-edged tool. From then on, men had an easy way of cutting meat, and even bone. They also used bone splinters for cutting up food and hides.

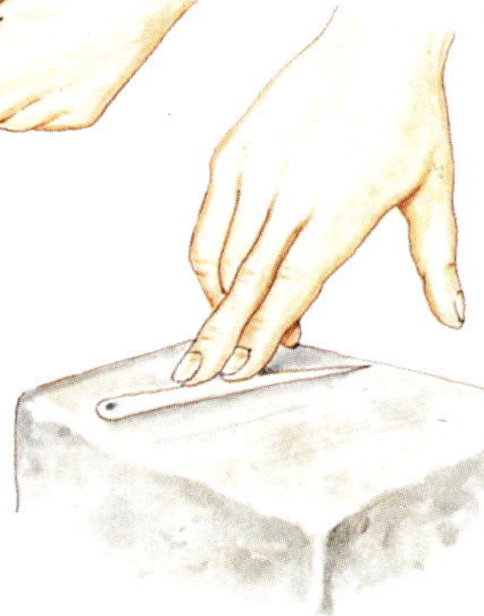

How to make a bone needle like those fashioned by ancient man.

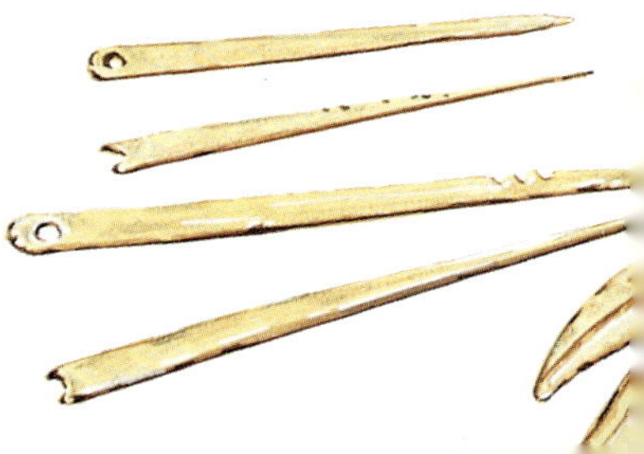

Toolmaking Progress

Early man must have been excited when he invented the first tool. And he must quickly have realized that if he could make one, he could make others. As long as 1.9 million years ago, *Homo habilis* was banging stones together to break and shape them into choppers, scrapers, and hammers. *Homo erectus,* about 1.4 million years ago, began making tools that had two sharp edges.

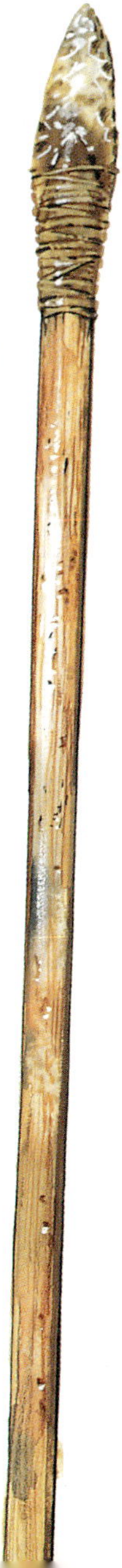

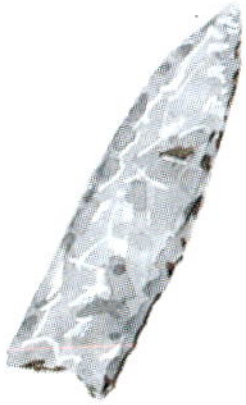

spear point

two-edged scraper

single-edged scraper

chopper

stone splinter

Toolmaking Takes Off

Cro-Magnons made tools for piercing holes in animal hides so they could more easily sew them together with the bone needles they also invented. They made special stone tools for carving bone, wood, and other stones. They learned how to shape bones into sharp points and fasten them to wooden shafts for spears and harpoons. It seems that whenever they saw a job that needed doing, they had the ability to make a tool to do it. At one site scientists discovered more than 50 different kinds of tools.

harpoons

The First Artists

Just like people of today, primitive humans felt the need to express themselves in various kinds of art.

Carving found in a cave in France.

Fragments of decorated bone.

Who Were the World's First Artists?

Neanderthals? Cro-Magnon? All we know is that ancient people placed a value on art. They carved detailed figures. On the walls of caves they drew pictures of animals and hunters. They also must have appreciated music, because ancient flute-like instruments have been found at some sites.

Ancient Engravings

Stone engravings as old as 35,000 years have been found in Europe. Those in Australia may be even older.

Relief
Raised carving.

The Talented Cro-Magnons

The Cro-Magnons may not have been the first artists, but they might have been the best. Cro-Magnons sculpted statuettes in bone and ivory, carved reliefs on stone walls, decorated their weapons with outlines of animal heads, and made jewelry from various kinds of gems. On the walls of the Lascaux Cave, in southern France, they painted superbly life-like figures of horses, bison, and other animals.

Painting Like a

The world's first artists got their paints from the earth and the many-colored rocks around them. They broke the rocks into pieces and hammered them into fine powders of red, yellow, brown, and black. By mixing clay or rock dust with water, they made paint. You will probably choose to use poster paint, but why not paint on rock, just as early man did?

Drawing with Charcoal

Use some charcoal left from your backyard barbecue. Primitive artists often saw ways to make bumps and cracks in a rock part of the drawing. Can you?

Painting with Your Fingers

Dip your fingertips in diluted paint for making points, marks, and parallel lines.

Painting with Swabs

Dip a ball of absorbent cotton or a piece of coarse cloth in paint diluted with water. Softly pat the paint onto the rock surface.

Primitive Man

To make a paintbrush

You will need:

- a small twig
- some string about 8 inches (20 cm) long

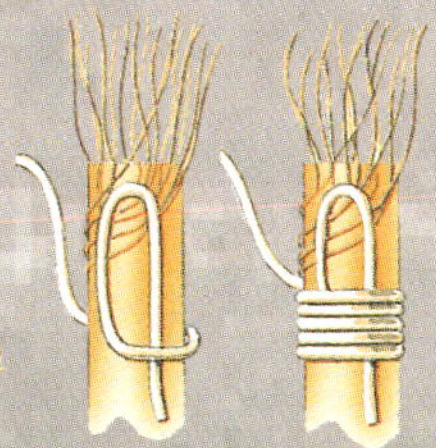

Fray the end of the string and attach it to one end of the twig, as shown in the illustration.

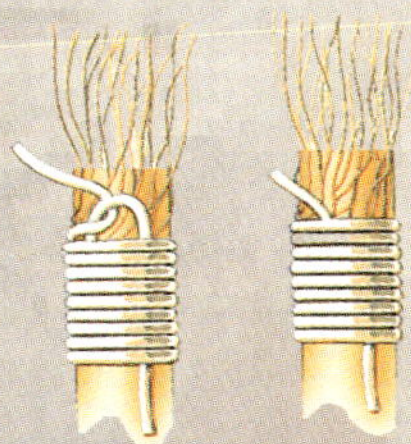

This is the way primitive people made the first paintbrushes.

Making a Negative Hand Print

Leave your mark, as many primitive people did, by placing your hand on a rock surface and then, through a straw, blowing paint around your fingers.

Engraving

Find a rock that is soft, such as sandstone, and draw a picture on it. With a nail or screwdriver, chip out the lines of your drawing.

Quiz

True or

All hominids disappeared 1 million years ago.

False. Most hominids died off, leaving man, who is a hominid, as the sole survivor.

Early humans had tails.

False. Early men were not monkeys. No hominid fossil showing evidence of a tail has ever been found.

Prehistoric man had a hard time finding enough food.

False. They found enough food to feed everyone, even those who didn't hunt—the craftsmen, the elderly, and the handicapped.

Cro-Magnons were the first people to have a language.

Probably false. The need to communicate by uttering sounds may go back to *Homo habilis*.

False?

■ **Humans had their beginnings in Africa.**

True, as far as we know. The oldest hominid fossils found so far were discovered in east Africa.

■ **There were once monkeys bigger than men.**

True. A race of huge monkeys called gigantopithecus lived in southeast Asia. They are believed to have died off 1 million years ago. Skeletons show them to have been taller than 9 feet (more than 3 meters).

■ **The Neanderthals had bigger brains than ours.**

True. All *Homo sapiens* of 50,000 years ago had bigger brains than do humans today.

■ **Prehistoric man was a cannibal.**

True. Primitive man often ate human flesh.

■ **Art is a Cro-Magnon invention.**

As far as we know, that is true. The earliest paintings yet discovered have been at Cro-Magnon sites.

True or

■ **Hominids became bipeds so they could see the dangers lurking in the savannah.**

False. The first hominids were barely 3 feet (about 1 meter) high—not tall enough to enable them to see over the savannah grasses.

■ **The first men were both hunters and scavengers.**

■ **Lucy ran in long bounding strides.**

False. Australopithecus was a biped but could not run like humans do today.

■ **Evolution stopped with the appearance of man.**

False. Evolution still continues. We can't know what man will be like a million years from now.

False?

True. They cut meat off carcasses they came upon but they also hunted small prey.

■ **All prehistoric men lived in caves.**

False. Caves were found only where there were mountains or rocky cliffs. Men lived mostly on the plains.

■ **Neanderthal people looked after those who were elderly or ill.**

True. They lived well enough to be able to take care of their weaker relatives.

■ **Cro-Magnon man wiped out the Neanderthals.**

False. They replaced them but very gradually. The process lasted 50,000 years in the Middle East, 10,000 years in Europe.

Index

Answers to the puzzle on pages 32–33.

)id you find all 19 items? They are: **dinosaur** *(dinosaurs were extinct before humans appeared),* **pteranodon** *(gone)efore there were men),* **baby bottle** *(nineteenth century),* **stamped letter** *(1820),* **tape measure** *(eighteenth century),* **;cissors** *(about 300 years ago),* **woman knitting** *(date of origin unknown, but it wasn't that long ago),* **watch** *(1675),* **'yeglasses** *(spectacles with side bows appeared in 1746),* **book** *(first books were printed in 1447),* **child's hoop** *'thought to be of Greek origin),* **lighter** *(nineteenth century),* **cooking pot** *(in use much later),* **sword** *(first century A.D.),* **)yramid** *(first built by Egyptians 4,000 years ago),* **field of corn** *(people had not yet developed agriculture),* **herd of :ows** *(people had not yet domesticated animals),* **tent pegs** *(twentieth century), and* **printing on the wall,** *"defense l'afficher," which is French for "Post no bills." (These people may have been living in what is now France but they ;ure didn't speak French, and they certainly didn't know how to write.)*

Photo credits for stickers
© M. Jelliffe/Ancient Art and Architecture Collection; Museum of Natural History, London; G. Blot/RMN; B. Wilson/Ancient Art and Architecture Collection; Museum of Natural History, London.

Photo credits for picture cards (left to right)
Top: © R. Sheridan/Ancient Art and Architecture Collection; Zefa; De Sazo/Rapho; © L. Bessol/M. N. H. N.;
Bottom: © P. van Gaalen; Horus-Zefa; Museum of Natural History, London; Zefa; F. Ducasse/Rapho; © Pratta R/Iconos.

Illustrations
Ian Chamberlain, Jeff Fisher, Chris Forsey, Daniel Guerrier, Christian Hook, Ruth Lindsay, Philippe Werner
Cover: Donald Grant

The title of the French edition is *À la rencontre des hommes préhistoriques*
Published by Les Editions Nathan, Paris

Barron's edition adapted by Carlisle Associates.

All inquiries should be addressed to:
Barron's Educational Series, Inc.
250 Wireless Boulevard
Hauppauge, NY 11788
Library of Congress Catalog Card No. 97-77433
International Standard Book No. 0-7641-5092-8
Printed in Italy
987654321

Stickers

The 2-million-year-old skull of an *Australopithecus robustus*, found in Tanzania, Africa.

Skull of "Taung's child." (See page 37.)

The cast of a Peking Man skull. (See page 36.)

Stone axe found in east Africa, dating from 1.7 million years ago.

Bifaced point made of flint about 50,000 B.C.

Scraping tool made and used about 40,000 B.C.

Sculpture of horse and bison carved about 20,000 B.C.

Leaf-shaped stone tool shaped about 40,000 B.C.

Stickers

Venus figure carved of mammoth ivory about 25,000 B.C.

Woman's head carved about 25,000 years B.C.

Engraved mammoth tusk dating from 25,000 to 30,000 B.C.

Head of a harpoon made perhaps 30,000 B.C.

Barbed point of mammoth ivory, carved between 18 and 30 thousand years B.C.

Bone needle made between 18 and 30 million years B.C.

Skull of a saber-toothed tiger that lived in North America 17,000 years ago.

Oil lamp found in France. Made between 13 and 20 thousand years ago.

Titles in the Megascope series:

Amazing Nature

Searching for Human Origins

Understanding the Human Body

Life in the Middle Ages

Mysteries, True and False

The Pharaohs of Ancient Egypt

BARRON'S

Barron's Educational Series, Inc.
250 Wireless Blvd., Hauppauge, NY 11788